The Bitcoin Revolution

By Dominick Barbato

Table of Contents

INTRODUCTION……………………………..
BASICS OF CRYPTOCURRENCIES…………..
BITCOINS FOR BUSINESS……………………
WHAT MAKES BITCOIN SO VOLATILE?….
BLOCKCHAIN TECHNOLOGY AND ITS IMPORTANCE…………………………………..
BEST CRYPTOCURRENCIES TO INVEST IN….
BITCOIN TRADING ONLINE………………
CONCLUSION…………………………………..

HAVE YOU EVER WANTED TO START A BUSINESS FROM THE COMFORT OF YOUR HOME?

HAVE YOU EVER TRIED TO START A BUSINESS BUT DIDN'T HAVE ALL THE RESOURCES?

DO YOU HAVE YOUR OWN BUSINESS BUT WANT TO SCALE IT AND DON'T KNOW HOW?

WELCOME TO....

ENHANCEDENTREPRENEURS.COM

THE #1 GLOBAL COMMUNITY FOR

ENTREPRENEURS LOOKING TO START, GROW AND TAKE THEIR BUSINESSES TO THE NEXT LEVEL!

ABOUT THE AUTHOR

Dominick Barbato is a 22-year-old serial entrepreneur, investor, mentor, business consultant, Youtuber and founder at Barbato Ventures, LLC. Dominick graduated college at 20 and earned his MBA in Organizational Leadership at 21. Currently, Dominick is apart of a few different companies in various industries. Dominick has started 18 companies, 9 of which "failed." Dominick continually looks to better himself by turning his mistakes into opportunities for growth.

Dominick now enjoys helping others

through various forms of content online to enhance thousands of lives daily! Dominick is an avid reader, writer, runner and world traveler. He loves sharing what he knows to anyone eager to learn. Currently residing in New York, Dominick has set the ambitious mission to help 100,000 people completely change the direction of their lives. Dominick believes you are as powerful as you believe you are.

Connect With Dominick:

www.dominickbarbato.com
www.enhancedentrepreneurs.com

Instagram: @thedombarbato
Twitter: @dominickbarbato
Facebook: Dominick Barbato

INTRODUCTION

In the times that we're living in, technology has made unbelievable advancements as compared to any time in the past. This evolution has redefined the life of man on almost every aspect. Moore's law states that our advances in technology now doubles every two years in processing power. Imagine the technology that will exist in the near future.

In fact, this evolution is an ongoing process and thus, human life on earth is improving constantly day in and day out.

One of the latest inclusions in this aspect is cryptocurrencies and the future of payment systems in the world.

Cryptocurrency is nothing but digital currency, which has been designed to impose security and anonymity in online monetary transactions. It uses cryptographic encryption to both generate currency and verify transactions.

The new coins are created by a process called mining, whereas the transactions are recorded in a public ledger, which is called the Transaction Block Chain.

Evolution of cryptocurrency is mainly attributed to the virtual world of the web and involves the procedure of transforming legible information into a code, which is almost uncrackable. Thus, it becomes easier to track purchases and transfers involving the currency.

Cryptography, since its introduction in the WWII to secure communication, has evolved in this digital age, blending with mathematical theories and computer science. Thus, it is now used to secure not only communication and information but also money transfers across the virtual web.

Over the past few years, people have been talking a lot about cryptocurrency. At first, this idea sounded scary but people started developing trust in it. You may have heard of Litecoin and Bitcoin. They both are cryptocurrencies that use the blockchain technology for highest security possible. Nowadays, these currencies are available in several different variations.

Understanding the latest version of technology, in form of cryptocurrency is not tough. You needs a little interest and a GUIDE like this to help you fully understand a new landscape of technology that is emerging.

CHAPTER 1

BASICS OF CRYPTOCURRENCIES

Cryptocurrency is digital money, which is designed in a way that it is secure and anonymous in some instances. It is closely associated with internet that makes use of cryptography, which is basically a process where legible information is converted into a code that cannot be cracked so as to tack all the transfers and purchases made.

Cryptography has a history dating back to the World War II, when there was a need to communicate in the most secure manner.

Since that time, an evolution of the same has occurred and it has become digitalized today where different elements of computer science and mathematical theory are being utilized for purposes of securing communications, money and information online.

The very first cryptocurrency was introduced in the year 2009 and is still well known all over the world. Many more cryptocurrencies have since been introduced over the past few years and today you can find so many available over the internet.

This kind of digital currency makes use of technology that is decentralized so as to allow the different users to make payments that are secure and also, to store money without necessarily using a name or even going through a financial institution. They are mainly run on a blockchain. A blockchain is a public ledger that is distributed publicly.

The cryptocurrency units are usually created using a process that is referred to as mining. This usually involves the use of a computer power. Doing it this way solves the math problems that can be very complicated in the generation of coins.

Users are only allowed to purchase the currencies from the brokers and then store them in cryptographic wallets where they can spend them with great ease.

Cryptocurrencies and the application of blockchain technology are still in the infant stages when thought of in financial terms. More uses may emerge in the future as there is no telling what else will be invented.

The future of transacting on stocks, bonds and other types of financial assets could very well be traded using the cryptocurrency and blockchain technology in the near future. Similar to credit cards, people need to slowly adopt and grow with technology. The landscape of technology is changing all over the world. The United States of America along with other currencies being printed across the globe are FIAT currencies. This is money backed by the government ledger. Essentially, if the government crashes your money is worthless. This is why you see some countries adopt bitcoin much more rapidly. Bitcoin and all coins developed on blockchain technology are based on a consensus currency, similar to gold. The more demand for it, the higher the price. Lets dig deeper into this.

Why use cryptocurrency?

One of the main traits of these currencies is the fact that they are secure and that they offer an anonymity level that you may not get anywhere else. There is no way in which a transaction can be reversed or faked. This is by far the greatest reason why you should consider using them.

The fees charged on this kind of currency are also quite low and this makes it a very reliable option when compared to the conventional currency. Since they are decentralized in nature, they can be accessed by anyone unlike banks where accounts are opened only by authorization.

Cryptocurrency markets are offering a brand new form of money and sometimes the rewards can be great. You may make a very small investment only to find that it has mushroomed into something great in a very short period of time. However, it is still important to note that the market can be volatile too, and there are risks that are associated with buying.

There is a level of anonymity associated with cryptocurrencies and this is a challenge because illegal activity can thrive here. This means that you need to be very careful when choosing to buy. Make sure you get your cryptocurrency from a trusted source.

How to use cryptocurrency

It is very easy for the ordinary people to make use of this digital currency. Just follow the steps given below:

You need a digital wallet (obviously, to store the currency)

Make use of the wallet to create unique public addresses (this enables you to receive the currency)

Use the public addresses to transfer funds in or out of the wallet

Cryptocurrency wallets

A cryptocurrency wallet is nothing else than a software program, which is capable to store both private and public keys. In addition to that, it can also interact with different blockchains, so that the users can send and receive digital currency and also keep a track on their balance.

The way the digital wallets work

In contrast to the conventional wallets that we carry in our pockets, digital wallets do not store currency. In fact, the concept of blockchain has been so smartly blended with cryptocurrency that the currencies never get stored at a particular location.

Nor do they exist anywhere in hard cash or physical form. Only the records of your transactions are stored in the blockchain and nothing else.

Suppose, a friend sends you some digital currency, say in form of bitcoin. What this friend does is he transfers the ownership of the coins to the address of your wallet. Now, when you want to use that money, you've unlock the fund.

In order to unlock the fund, you need to match the private key in your wallet with the public address that the coins are assigned to. Only when both these private and public addresses match, your account will be credited and the balance in your wallet will swell.

Simultaneously, the balance of the sender of the digital currency will decrease. In transactions related to digital currency, the actual exchange of physical coins never take place at any instance.

Understanding the cryptocurrency address

By nature, it is a public address with a unique string of characters. This enables a user or owner of a digital wallet to receive cryptocurrency from others. Each public address, that is generated, has a matching private address.

This automatic match proves or establishes the ownership of a public address. As a more practical analogy, you may consider a public cryptocurrency address as your email address to which others can send emails. The emails are the currency that people send you.

How can cryptocurrency help you?

As far as fraud is concerned, this type of currency can't be faked as it's in digital form and can't be reversed or counterfeited unlike the credit cards.

Immediate settlement

Buying real property involves third parties, such as lawyers and notary. So, delays can occur and extra costs may incur. On the other hand, bitcoin contracts are designed and enforced in order to include or exclude third parties. The transactions are quick and settlements can be made instantly.

Lower fees

Typically, there is no transaction fee if you want to exchange bitcoin or any other currency. For verifying a transaction, there are minors who get paid by the network.

Although there is zero transaction fee, most buyers or sellers hire the services of a third-party, such as coinbase for the creation and maintenance of their wallets. If you don't know, these services function just like paypal that offers a web-based exchange system.

Identification of theft

Your merchant gets your full credit line when you provide them with your credit card. This is true even if the transaction amount is very small. Actually, what happens is that credit cards work based on a "pull" system where the online store pulls the required amount from the account associated with the card.

On the other hand, the digital currencies feature a "push" mechanism where the account holder sends only the amount required without any additional information. So, there is no chance of theft.

Open access

According to statistics, there are around 2.2 billion people who use the Internet but not all of them have access to the conventional exchange. So, they can use the new form of payment method.

Decentralization

As far as decentralization is concerned, an international computer network called blockchain technology manages the database of bitcoin. In other words, bitcoin is under the administration of the network, and there is no central authority.

In other words, the network works on a peer-to-peer based approach.

Recognition

Since cryptocurrency is not based on the exchange rates, transaction charges or interest rates, you can use it internationally without suffering from any problems. So, you can save a lot of time and money. In other words, bitcoin and other currencies like this are recognized all over the world. You can count on them.

So, if you have been looking for a way to invest your extra money, you can consider investing in bitcoin. You can either become a miner or investor. However, make sure you know what you are doing.

Safety is not an issue but other things are important to be kept in mind.

CHAPTER 2

BITCOINS FOR BUSINESS

The cryptocurrency that continues to mesmerize the world, the first of its kind, bitcoin was once entirely a classy realm of tech-geniuses who were keen to uphold the philosophy of maximizing autonomy, but bitcoin has a shot to fame with the promise of a wide consumer base.

Yet, to the uninitiated consumers, a query remains. Some are really yet to unravel this overly fluctuating cryptocurrency. Generated and stored electronically, bitcoin is actually a form of digital currency. The network can't actually be controlled by anyone, the currency is decentralized.

It's not an actual coin, it's "cryptocurrency," a digital form of payment that is produced ("mined") by lots of people worldwide. It allows peer-to-peer transactions instantly, worldwide, for free or at very low cost.

Bitcoin was invented after decades of research into cryptography by software developer, Satoshi Nakamoto (believed to be a pseudonym), who designed the algorithm and introduced it in 2009. His true identity remains a mystery. People believe Satoshi to be a team, but its all rumors and minimal fact.

This currency is not backed by a tangible commodity (such as gold or silver); bitcoins are traded online which makes them a commodity in themselves. Bitcoin is an open-source product, accessible by anyone who is a user. Bitcoin does trade against the USD in some cases though, like when using an online platform like Binance. All you need is an email address, Internet access, and money to get started.

Where does it come from?

Bitcoin is mined on a distributed computer network of users running specialized software; the network solves certain mathematical proofs, and searches for a particular data sequence ("block") that produces a particular pattern when the BTC algorithm is applied to it. A match produces a bitcoin. It's complex and time- and energy-consuming.

Only around 21 million bitcoins are ever to be mined (about 11 million are currently in circulation). The math problems the network computers solve get progressively more difficult to keep the mining operations and supply in check.

This network also validates all the transactions through cryptography.

How does bitcoin work?

Internet users transfer digital assets (bits) to each other on a network. There is no online bank; rather, bitcoin has been described as an Internet-wide distributed ledger. Users buy bitcoin with cash or by selling a product or service for bitcoin.

Bitcoin wallets store and use this digital currency. Users may sell out of this virtual ledger by trading their bitcoin to someone else who wants in. Anyone can do this, anywhere in the world. There are smartphone apps for conducting mobile bitcoin transactions and bitcoin exchanges are populating the Internet.

How is bitcoin valued?

Bitcoin is not held or controlled by a financial institution; it is completely decentralized. Unlike real-world money it cannot be devalued by governments or banks.

Instead, bitcoin's value lies simply in its acceptance between users as a form of payment and because its supply is finite. Its global currency values fluctuate according to supply and demand and market speculation; as more people create wallets and hold and spend bitcoins, and more businesses accept it, bitcoin's value will rise.

Banks are now trying to value bitcoin and some investment funds are now even offering bitcoin going into 2018.

What are its benefits?

There are benefits to consumers and merchants that want to use this payment option.

1. Fast transactions - Bitcoin is transferred instantly over the Internet.

2. No fees/low fees -- Unlike credit cards, bitcoin can be used for free or very low fees. Without the centralized institution as middle man, there are no authorizations

(and fees) required. This improves profit margins sales.

3. Eliminates fraud risk -Only the bitcoin owner can send payment to the intended recipient, who is the only one who can receive it. The network knows the transfer has occurred and transactions are validated; they cannot be challenged or taken back.

This is big for online merchants who are often subject to credit card processors' assessments of whether or not a transaction is fraudulent, or businesses that pay the high price of credit card chargebacks.

4. Data is secure -- As we have seen with recent hacks on national retailers' payment processing systems, the Internet is not always a secure place for private data. With bitcoin, users do not give up private information.

a. They have two keys - a public key that serves as the bitcoin address and a private key with personal data.

b. Transactions are "signed" digitally by combining the public and private keys; a mathematical function is applied and a certificate is generated proving the user initiated the transaction. Digital signatures are unique to each transaction and cannot be re-used.

c. The merchant/recipient never sees your secret information (name, number, physical address) so it's somewhat anonymous but it is traceable (to the bitcoin address on the public key).

5. Convenient payment system -- Merchants can use bitcoin entirely as a payment system; they do not have to hold any bitcoin currency since bitcoin can be converted to dollars. Consumers or merchants can trade in and out of bitcoin and other currencies at any time.

6. International payments - Bitcoin is used around the world; e-commerce merchants and service providers can easily accept

international payments, which open up new potential marketplaces for them.

7. Easy to track -- The network tracks and permanently logs every transaction in the bitcoin block chain (the database). In the case of possible wrongdoing, it is easier for law enforcement officials to trace these transactions.

8. Micropayments are possible - Bitcoins can be divided down to one one-hundred-millionth, so running small payments of a dollar or less becomes a free or near-free transaction. This could be a real boon for convenience stores, coffee shops, and subscription-based websites (videos, publications).

Bitcoin Transactions

Bitcoin in the retail environment

At checkout, the payer uses a smartphone app to scan a QR code with all the transaction information needed to transfer the bitcoin to the retailer. Tapping the "Confirm" button completes the transaction. If the user doesn't own any bitcoin, the network converts dollars in his account into the digital currency.

The retailer can convert that bitcoin into dollars if it wants to, there were no or very low processing fees (instead of 2 to 3 percent), no hackers can steal personal consumer information, and there is no risk of fraud. The cost of keeping the bitcoin network together is increasing therefor processing fees is also increasing.

Bitcoins in hospitality

Hotels can accept bitcoin for room and dining payments on the premises for guests who wish to pay by bitcoin using their mobile wallets, or PC-to-website to pay for a reservation online. A third-party BTC merchant processor can assist in handling the transactions which it clears over the bitcoin network.

These processing clients are installed on tablets at the establishments' front desk or in the restaurants for users with BTC smartphone apps. (These payment processors are also available for desktops, in retail POS systems, and integrated into foodservice POS systems.) No credit cards or money need to change hands.

These cashless transactions are fast and the processor can convert bitcoins into currency and make a daily direct deposit into the establishment's bank account. It was announced in January 2014 that two Las Vegas hotel-casinos would accept bitcoin payments at the front desk, in their restaurants, and in the gift shop.

Business owners should consider issues of participation, security and cost.

• A relatively small number of ordinary consumers and merchants currently use or understand bitcoin. However, adoption is increasing globally and tools and technologies are being developed to make participation easier.

- It's the Internet, so hackers are threats to the exchanges. The Economist reported that a bitcoin exchange was hacked in September 2013 and $250,000 in bitcoins was stolen from users' online vaults. Bitcoins can be stolen like other currency, so vigilant network, server and database security is paramount. Unless you keep your crypto in a physical crypto wallet. These wallets act as security on top of where you hold your bitcoin.

- Users must carefully safeguard their bitcoin wallets which contain their private keys. Secure backups or printouts are crucial.

- Bitcoin is not regulated or insured by the US government so there is no insurance for your account if the exchange goes out of business or is robbed by hackers.

- Bitcoins are relatively expensive. Current rates and selling prices are available on the online exchanges.

The virtual currency is not yet universal but it is gaining market awareness and acceptance. A business may decide to try bitcoin to save on credit card and bank fees, as a customer convenience, or to see if it helps or hinders sales and profitability.

With numerous enthusiasts who are keen to trade bitcoins, the young currency and all the craze surrounding it seems to grow a little bit every day. All the knowledge associated with it seems to be as important as the currency itself.

The significance of a "Bitcoin wiki", an autonomous project, cannot be denied at all. It will act as a storehouse of knowledge for bitcoin enthusiasts all around the world.

CHAPTER 3

WHAT MAKES BITCOIN SO VOLATILE?

Traders are always concerned about bitcoin''s volatility. It is important to know what makes the value of this particular digital currency highly unstable. Just like many other things, the value of bitcoin also depends upon the rules of demand and supply.

If the demand for bitcoin increases, then the price will also increase. On the contrary side, the decrease in demand for the bitcoin will lead to decreased demand.

In simple words, we can say that the price is determined by what amount the trading market is agreed to pay. If a large number of people wish to purchase bitcoins, then the price will rise. If more folks want to sell bitcoins, then the price will come down.

It is worth knowing that the value of bitcoin can be volatile if compared to more established commodities and currencies.

This fact can be credited to its comparatively small market size, which means that a lesser amount of money can shift the price of bitcoin more prominently. This inconsistency will reduce naturally over the passage of time as the currency develops and the market size grows.

After being teased in late 2016, bitcoin touched a new record high level in the first week of the current year. There could be several factors causing the bitcoin to be volatile. Some of these are discussed here. As of 2018, the price of bitcoin is fluctuating between $14,000 and $17,000 which is substantially higher.

Press Factor

Bitcoin users are mostly scared by different news events including the statements by government officials and geopolitical events that bitcoin can be possibly regulated. It means the rate of bitcoin adoption is troubled by negative or bad press reports.

Different bad news stories created fear in investors and prohibited them from investing in this digital currency.

An example of bad headline news is the eminent utilization of bitcoin in processing drug transactions through Silk Road which came to an end with the FBI stoppage of the market in October 2013. This sort of stories produced panic among people and caused the bitcoin value to decrease greatly.

On the other side, veterans in the trading industry saw such negative incidents as an evidence that the bitcoin industry is maturing. So the bitcoin started to gain its increased value soon after the effect of bad press vanished.

Fluctuations of the Perceived Value

Another great reason for bitcoin value to become volatile is the fluctuation of the bitcoin's perceived value. You may know that this digital currency has properties akin to gold. This is ruled by a design decision by the makers of the core technology to restrict its production to a static amount, 21 million BTC. Due to this factor, investors may allocate less or more assets in into bitcoin.

News about Security Breaches

Various news agencies and digital media play an important role in building a negative or positive public concept. If you see something being advertised advantageously, you are likely to go for that without paying much attention to negative sides.

There has been news about bitcoin security breaches and it really made the investors think twice before investing their hard earned money in bitcoin trading. They become too susceptible about choosing any specific bitcoin investment platform.

'Bitcoin' may become volatile when bitcoin community uncovers security susceptibilities in an effort to create a great open source response in form of security fixes.

Such security concerns give birth to several open-source software such as

Linux. Therefore, it is advisable that bitcoin developers should expose security vulnerabilities to the general public in order to make strong solutions.

The 'OpenSSL' weaknesses was attacked by 'Heartbleed' bug and reported by Neel Mehta (a member of Google's security team) on April 1, 2014, appear to had some descending effect on the value of bitcoin. According to some reports, the bitcoin value decreased up to 10% in the ensuing month as compared to the U.S. Dollar.

Small option value for holders of large bitcoin Proportions

The volatility of bitcoin also depends upon bitcoin holders having large proportions of this digital currency. It is not clear for bitcoin investors (with current holdings over $10M) that how they would settle a position that expands into a fiat position without moving the market severely.

So 'bitcoin has not touched the bulk market adoption rates that would be important to give option value to large bitcoin holders.

All these losses and the resultant news about heavy losses had a dual effect on instability. You may not know that this reduced the general float of bitcoin by almost 5%. This also created a potential lift on the residual bitcoin value due to the reason of increased scarcity.

Nevertheless, superseding this lift was the negative outcome of the news series that followed. Particularly, many other bitcoin gateways saw the large failure at Mt Gox as an optimistic thing for the long-term prospects of the bitcoin.

CHAPTER 4

BLOCKCHAIN TECHNOLOGY AND ITS IMPORTANCE

Blockchain is an irrefutably resourceful invention which is practically bringing about a revolution in the global business market. Its evolution has brought with it a greater good, not only for businesses but for its beneficiaries as well. But since it's revelation to the world, a vision of its operational activities is still unclear.

We must keep in mind that bitcoin is setting a foundation for more companies to develop on top of blockchain technology. The future of blockchain technology is bright, along with bitcoin acting as the first currency to get rapidly adopted by an entire population.

Blockchain technology serves as a platform that allows the transit of digital information without the risk of being copied. It has, in a way, laid the foundation of a strong backbone of a new kind of internet space.

Originally designed to deal with bitcoin, trying to explain the layman about the functions of its algorithms, the hash functions, and digital signature property, today, the technology buffs are finding other potential uses of this immaculate invention which could pave the way to the onset of an entirely new business dealing process in the world.

Blockchain, to define in all respects, is a kind of algorithm and data distribution structure for the management of electronic cash without the intervention of any centralized administration, programmed to record all the financial transactions as well as everything that holds value.

The Working of Blockchain

Blockchain can be comprehended as Distributed Ledger technology which was originally devised to support the bitcoin cryptocurrency. But post heavy criticism and rejection, the technology was revised for use in things more productive.

To give a clear picture, imagine a spreadsheet that's practically augmented tons to times across a plethora of computing systems. And then imagine that these networks are designed to update this spreadsheet from time to time. This is exactly what blockchain is.

Information that's stored on a blockchain is a shared sheet whose data is reconciled from time to time. It's a practical way that speaks of many obvious benefits. To being with, the blockchain data doesn't exist in one single place.

This means that everything stored in there is open for public view and verification. Further, there isn't any centralized information storing platform which hackers can corrupt. It's practically accessed over a million computing systems side-by-side, and its data can be consulted by any individual with an internet connection.

Durability and Authenticity of Blockchain

Blockchain technology is something that minims the internet space. It's chic robust in nature. Similar to offering data to the general public through the World Wide Web, blocks of authentic information are stored on blockchain platform which is identically visible on all networks.

Vital to note, blockchain cannot be controlled by a single people, entity or identity, and has no one point of failure. Just like the internet has proven itself as a durable space since last 30 years, blockchain too will serve as an authentic, reliable global stage for business transaction as it continues to develop.

Transparency and Incorruptible Nature

Veterans of the industry claim that blockchain lives in a state of consciousness. It practically checks on itself every now and then. It's similar to a self-auditing technology where its network reconciles every transaction, known as a block, which happens aboard at regular intervals.

This gives birth to two major properties of blockchain - it's highly transparent, and at the same time, it cannot be corrupted. Each and every transaction that takes place on this server is embedded within the network, hence, making the entire thing very much visible all the time to the public.

Furthermore, to edit or omit information on blockchain asks for a humongous amount of efforts and a strong computing power. Amid this, frauds can be easily identified. Hence, it's termed incorruptible.

Users of Blockchain

There isn't a defined rule or regulation about who shall or can make use of this immaculate technology. Though at present, its potential users are banks, commercial giants and global economies only, the technology is open for the day to day transactions of the general public as well. The only drawback blockchain is facing is global acceptance.

CHAPTER 5

BEST CRYPTOCURRENCIES TO INVEST IN

Whether it's the idea of cryptocurrencies itself or diversification of their portfolio, people from all walks of life are investing in digital currencies. If you're new to the concept and wondering what's going on, here are some basic concepts and considerations for investment in cryptocurrencies.

An important consideration is storage of the coins. One option, of course, is to store it on the exchange where you buy them. However, you will have to be careful in selecting the exchange.

The popularity of digital currencies has resulted in many new, unknown exchanges popping up everywhere. Take the time to do your due diligence so you can avoid the scammers.

Another option you have with cryptocurrencies is that you can store them yourself. One of the safest options for storing your investment is hardware wallets. Companies like Ledger allow you store bitcoins and several other digital currencies as well.

What's the market like and how can I learn more about it?

The cryptocurrency market fluctuates a lot. The volatile nature of the market makes it more suited for a long-term play. Digital currencies aim to disrupt the traditional currency and commodity market.

While these currencies still have a long way to go, the success of bitcoins and ethereum have proven that there is genuine interest in the concept. Understanding the basics of cryptocurrency investment will help you start in the right way.

In the previous year the value of Bitcoin has soared. There are also new cryptocurrencies on the market, which is even more surprising which brings cryptocoins' worth up to more than one hundred billion.

On the other hand, the longer term cryptocurrency-outlook is somewhat of a blur. There are squabbles of lack of progress among its core developers which make it less alluring as a long term investment and as a system of payment.

Bitcoin

Still the most popular, bitcoin is the cryptocurrency that started all of it. Being first to market, there are a lot of exchanges for bitcoin trade all over the world. BitStamp and Coinbase are two well-known US-based exchanges.

Coinbase is an established American exchange. If you are interested in trading other digital currencies along with bitcoin, then a crypto marketplace is where you will find all the digital currencies in one place.

Many exchanges are actually starting to accept less and less people or even completely disabling sign-ups due to the inflow

Both as a payment system and as a stored value, bitcoin enables users to easily receive and send bitcoins. The concept of the blockchain is the basis in which bitcoin is based. It is necessary to understand the blockchain concept to get a sense of what the cryptocurrencies are all about.

Litecoin

One alternative to bitcoin, litecoin attempts to resolve many of the issues that hold bitcoin down. It is not quite as resilient as ethereum with its value derived mostly from adoption of solid users. It pays to note that Charlie Lee, ex-Googler leads litecoin. He is also practicing transparency with what he is doing with Litecoin and is quite active on Twitter.

Litecoin was bitcoin's second fiddle for quite some time but things started changing early in the year of 2017. First, litecoin was adopted by coinbase along with ethereum and bitcoin. Next, litecoin fixed the bitcoin issue by adopting the technology of Segregated Witness. This gave it the capacity to lower transaction fees and do more.

Ethereum

Vitalik Buterin, superstar programmer thought up ethereum, which can do everything bitcoin is able to do. However its purpose, primarily, is to be a platform to build decentralized applications. The blockchains are where the differences between the two lie.

Basically, the blockchain of Bitcoin records a contract-type, one that states whether funds have been moved from one digital address to another address. However, there is significant expansion with Ethereum as it has a more advanced language script and has a more complex, broader scope of applications.

Monero

Monero aims to solve the issue of anonymous transactions. Even if this currency was perceived to be a method of laundering money, monero aims to change this.

Basically, the difference between monero and bitcoin is that bitcoin features a transparent blockchain with every transaction public and recorded. With bitcoin, anyone can see how and where the money was moved.

There is some somewhat imperfect anonymity on bitcoin, however. In contrast, monero has an opaque rather than transparent transaction method. No one is quite sold on this method but since some folks love privacy for whatever purpose, monero is here to stay.

Zcash

Not unlike monero, zcash also aims to solve the issues that bitcoin has. The difference is that rather than being completely transparent, monero is only partially public in its blockchain style.

Zcash also aims to solve the problem of anonymous transactions. After all, no every person loves showing how much money they actually spent on memorabilia by Star Wars.

Thus, the conclusion is that this type of cryptocoin really does have an audience and a demand, although it's hard to point out which cryptocurrency that focuses on privacy will eventually come out on top of the pile.

Ripple

Ripple is a centralized coin. This means it not mineable and all the coins are stored within one place. Ripple is a real-time settlement system called the Ripple protocol. Ripple has recently received a lot of hype due to what they are trying to accomplish. They also have the quickest transaction times and lowest transactions fee. This coin grew over 35,000% in under a year, and 1,200% in a year.

EOS

Another competitor of Ethereum, EOS promises to solve the scaling issue of ethereum through the provision of a set of tools that are more robust to run and create apps on the platform.

It is incredibly hard to predict which bitcoin in the list will become the next superstar. However, user adoption has always be one key success factor when it came to cryptocurrencies.

Both ethereum and bitcoin have this and even if there is a lot of support from early adopters of every cryptocurrency in the list, some have yet to prove their staying power. Nonetheless, these are the ones to invest in and watch out for in the coming months.

CHAPTER 6

BITCOIN TRADING ONLINE

Bitcoin is a peer-to-peer payment system, otherwise known as electronic money or virtual currency. It offers a twenty-first century alternative to brick and mortar banking. Exchanges are made via "e wallet software". The bitcoin has actually subverted the traditional banking system, while operating outside of government regulations.

Bitcoin uses state-of-the-art cryptography, can be issued in any fractional denomination, and has a decentralized distribution system, is in high demand globally and offers several distinct advantages over other currencies such as the US dollar. For one, it can never be garnished or frozen by the bank(s) or a government agency.

Back in 2009, when the bitcoin was worth just ten cents per coin, you would have turned a thousand dollars into millions, if you waited just eight years. The number of bitcoins available to be purchased is limited to 21,000,000.

According to Bill Gates, "Bitcoin is exciting and better than currency". Bitcoin is a decentralized form of currency. There is no longer any need to have a "trusted, third-party" involved with any transactions.

By taking the banks out of the equation, you are also eliminating the lion's share of each transaction fee. In addition, the amount of time required to move money from point A to point B, is reduced formidably.

The largest transaction to ever take place using bitcoin is one hundred and fifty million dollars. This transaction took place in seconds with minimal fee's.

In order to transfer large sums of money using a "trusted third-party", it would take days and cost hundreds if not thousands of dollars. This explains why the banks are violently opposed to people buying, selling, trading, transferring and spending bitcoins.

Only .003% of the worlds (250,000) population is estimated to hold at least one bitcoin. And only 24% of the population know what it is. Bitcoin transactions are entered chronologically in a 'blockchain' just the way bank transactions are.

Blocks, meanwhile, are like individual bank statements. In other words, blockchain is a public ledger of all Bitcoin transactions that have ever been executed. It is constantly growing as 'completed' blocks are added to it with a new set of recordings. To use conventional banking as an analogy, the blockchain is like a full history of banking transactions.

Setting Up Your E Wallet Software Account

As soon as you create your own unique e wallet software account, you will have the ability to transfer funds from your e wallet to a recipients e wallet, in the form of bitcoin.

To facilitate the transfer of your funds in bitcoin to and from a trading platform, you will simply link your e wallet 'address' to the e wallet 'address' of your chosen trading platform. In actuality, it is much easier than it sounds. The learning curve in relation to using your e wallet, is very short.

To set up an e wallet, there are a myriad of company's online that offer safe, secure, free and turn-key e-wallet solutions. Coinbase, Binance, Bitstamp are the top three for purchasing cryptocurrencies. A simple Google search will help you find the right e wallet software for you, depending upon what your needs are.

Many people get started using a "blockchain" account. This is free to set up and very secure. You have the option of setting up a two-tier login protocol, to further enhance the safety and security, in relation to your e wallet account, essentially protecting your account from being hacked into.

There are many options when it comes to setting up your e wallet. A good place to start is Coinbase. You can find them by doing a Google search. Furthermore, bitcoins that are funded in Coinable are safe.

To withdraw money is just as easy as funding your Coinbase account. You just go to "buy/sell" and sell whatever quantity you wish or buy whatever quantity you wish.

Purchase Any Fractional Denomination Of Bitcoin

To buy any amount of bitcoin, you are required to deal with a digital currency broker. As with any currency broker, you will have to pay the broker a fee, when you purchase your bitcoin.

It is possible to buy .1 of of bitcoin or less if that is all that you would like to purchase. The cost is simply based on the current market value of a full bitcoin at any given time.

There are a myriad of bitcoin brokers online. A simple Google search will allow you to easily source out the best one for you. It is always a good idea to compare their rates prior to proceeding with a purchase. You should also confirm the rate of a bitcoin online, prior to making a purchase through a broker, as the rate does tend to fluctuate frequently.

Stay Away From Any Trading Platfrom Promising Unrealistic Returns To Unsuspecting Investors

Finding a reputable bitcoin trading company that offers a high return is paramount to your online success. Earning 1% per day is considered a high return in this industry. Earning 10% per day is impossible. With online bitcoin trading, it is feasible to double your digital currency within ninety days.

You must avoid being lured by any company that is offering returns such as 10% per day. This type of a return is not realistic with digital currency trading. There was a company called Coinexpro that was offering 10% per day to bitcoin traders.

And it ended up being a ponzi scheme. If it's 10% per day, walk away. The aforementioned trading platform appeared to be very sophisticated and came across as being legitimate. My advice is to focus on trading your bitcoin with a company that offers reasonable returns such as 1% per day.

There will be other companies that will attempt to separate you from your bitcoin using unscrupulous methods. Be very cautious when it comes to any company that is offering unrealistic returns. Once you transfer your bitcoin to a recipient, there is literally nothing your can do to get it back. You must ensure that your chosen trading company is fully automated & integrated with blockchain, from receipt to payment.

More importantly, it is crucial that you learn to differentiate legitimate trading opportunities from unscrupulous "company's" that are experts when it comes to separating it's clients from their money.

The bitcoin and other digital currencies are not the issue. It is the trading platforms that you must exercise caution with, prior to handing over your hard-earned money.

Your ROI should also be upwards of 1%+ per day because the trading company that you are lending your bitcoin to, is most likely earning upwards of 5%+ per day, on average. Your ROI must also be automatically transferred into your "e-wallet" at regular intervals, throughout your contract term.

CONCLUSION

Cryptocurrency is the newest trend in the money market that contains the elements of computer science and mathematical theory.

Its primary function is to secure communication as it converts legible information into an unbreakable code. You can track your purchases and transfers with cryptocurrency. Following are the top ten tips for investors to invest in cryptocurrency.

Investing in cryptocurrency is just like investing in any other commodity. It has two faces - it can be used as an asset or as an investment, which you can sell and exchange.

Buy bitcoins directly if you do not want to pay the fee for investing or if you are interested in possessing real bitcoins. There are a lot of options all over the world including Bitcoin.de, BitFinex, and BitFlyer from where you can buy bitcoins directly.

Today, bitcoin is the most common cryptocurrency in the world of investment. In the United States, only 24% of the adults know about it, and surprisingly only 4% Americans use it. It is good news for the financial investors as the low usage represents a fruitful investment for the future.

The combined market cap of the cryptocurrencies is more than 700 billion American dollars. It includes all cryptocurrencies in existence including hundreds of smaller and unknown ones. The real-time usage of the cryptocurrencies has gone up, showing a rise in trend.

As an investor, the usage must be the key for you. The demand and supply data of cryptocurrencies exhibits a decent investment opportunity right now. There exists a strong usage of the currencies for facilitating payments between financial institutions and thus, pushing transaction costs down meaningfully.

Currently, the cryptocurrency market is in euphoria. It is the point where the investment may not appear as a golden opportunity to you but the values will go higher from here. Businesses, governments, and society across the globe will soon be considering cryptocurrencies.

www.ingramcontent.com/pod-product-compliance
Lightning Source LLC
Chambersburg PA
CBHW030443220526
45464CB00006B/2404